Community Health Workers of Pakistan & Their Attrition

Inferences of a Survey

Romana Ayub

&

Saad Siddiqui

Copyright © 2013 Lulu Author.

All rights reserved.

No part of this publication may be reproduced, distributed, or transmitted in any form or by any means, including photocopying, recording, or other electronic or mechanical methods, without the prior written permission of the publisher, except in the case of brief quotations embodied in critical reviews and certain other noncommercial uses permitted by copyright law.

Publisher: Lulu Press, Inc

Raleigh, North Carolina. USA

Printed in the USA

ISBN: 978-1-304-51293-2

Contributors

1. Dr. Romana Ayub
MBBS (Peshawar), MPH (Pakistan), MSc International Health (Germany)
Assistant Professor, Department of Community Medicine
Khyber Medical College, Peshawar, Pakistan.

2. Dr. Saad Siddiqui
MBBS (AKU)
Intern, The Aga Khan University Medical College,
Karachi, Pakistan.

3. Dr. Rainer Kuelker
Department of Tropical Hygiene and Public Health,
University of Heidelberg, Germany

4. Dr. Inamullah
UNICEF, Pakistan.

Dedications

Dedicated to my beloved Mother. We will always remember her with profound love and respect.

Romana Ayub
November 2012

I would like to dedicate this work to my Late Grandparents Mr. & Mrs. S. Chiragh Shah for their profound affection & being a source of inspiration.

Saad Siddiqui

November 2012

Contents

Preface

Précis

Abbreviations

1. Prologue
   1.1 Background
   1.2 Lady health worker of NP-FPPHC Pakistan

2. Community Health Worker programmes around the world
   2.1 General information of Human Resource situation In different countries
   2.2 General Information on Volunteer/ Community Based programmes
   2.3 The Problem of Attrition in VHW programmes
   2.4 Examples from around the globe
   2.5 The LHW Programme in Pakistan
   2.6 Other causes of Attrition in Pakistan

3. The Survey
   3.1 Design
   3.2 Data collection & Survey methodology
   3.3 Population & Sampling
   3.4 Analysis
   3.5 Limitations
   3.6 Ethical considerations

4. Inferences

5. Argument

    5.1   Importance of Knowledge

    5.2   Domestic reasons Affecting Attrition

    5.3   Attrition due to lack of Incentives

    5.4   Health vs. Attrition

    5.5   Behavioral Factors

    5.6   Importance of Community Opinion

    5.7   Self perception of LHWs

    5.8   Motivating vs. discouraging factors

    5.9   Problems related to training

    5.10   Aspects related to job satisfaction

    5.11   Problems related to male supervisory visits

    5.12   Importance of Attrition

6. Key Recommendations

    6.1   Policy formulation

    6.2   Enhancement of Salary

    6.3   Medical Endowment

    6.4   Formulation of Incentives

    6.5   Behavioural Change

    6.6   Improvement of Records

    6.7   Creating impact through ownership

    6.8   Monitoring of training

References

Acknowledgements

Annexes

    Annex 1: Map of Pakistan showing Khyber Pakhtunkhwa
    Annex 2: Survey Districts Profile
    Annex 3: Document Review Tool
    Annex 4: Ethical Note

Preface

Primary Health care along with all its components is a neglected area in Health system of Pakistan. Majority of investments are concentrated to development of tertiary care to neglect of Primary & secondary level services.

Pakistan overall faces a shortage of health professionals but the main problem lies in concentration of existing health personnel in major towns, leaving periphery completely neglected. This shortage exists at levels in hierarchy of health professionals. By this work we made an effort to highlight problems faced by Community health workers of Pakistan. These workers are not only an efficient way of delivery of Primary healthcare services but as they cover the periphery well, they also improve overall coverage of healthcare.

Compared to other countries in region Pakistan is relatively performing well as far as indices relating to community healthcare professionals are concerned. By bringing about a change in policies, this already good working programme can be strengthened, eventually adding to betterment of society.

Romana Ayub & Saad Siddiqui

November, 2012.

Précis

This work is based upon a survey that was conducted to find out causes of attrition in community healthcare professionals in Pakistan, with model being Lady Health workers programme in two Districts of Khyber Pakhtunkhwa (i.e. Mardan & Nowshera).

The present security situation has adversely affected functioning of all sectors, but National Programme for Family Planning and Primary Health Care (NP-FPPHC) is one of the obvious victims; due to its female employees and fieldwork involved as part of the job. The records of this program show that districts of Mardan and Nowshera have a well established Lady Health Workers programme. Reasons for this being efficient programme managers and increased inputs from partner organizations which bringing about significant improvements in functioning of programme.

The main objective of this work was to highlight the causes of attrition and to find out different trends in attrition. After reviewing the data provided by both districts, the attrition rate in Mardan for year 2008 was found to be (4.2%), whereas for Nowshera it was (5.4%).

The core reasons of attrition inferred from findings of this survey were;

- Domestic reasons like marriage, less time for children, and transfer of husband.
- Inadequate salary and too much work load (mostly due to National Immunization days).
- Ill health, uncooperative behavior of the Lady Health Supervisor and perception of poor respect by the community were other reported

problems.

Moreover the survey findings also revealed that the programme records were not being well maintained in some of the districts. Some discrepancy was also noticed in flow of information from the district to the provincial level which resulted in misleading information. To help overcome the issues reported, there is need to review policies, especially those governing recruitment of healthcare professionals.

ABBREVIATIONS AND ACRONYMS

APC	Assistant Provincial Coordinator
ADC	Assistant District Coordinator
BHU	Basic Health Unit
BRAC	Bangladesh Rural Advancement Committee
CDD	Community Directed Distributers
CPR	Contraceptive Prevalence Rate
CHW	Community Health Worker
DC	District Coordinator
EPI	Expanded Programme on Immunization
EDOH	Executive District Officer Health
GOP	Government of Pakistan
HMIS	Health Management Information System
LHS	Lady Health Supervisor
LHV	Lady Health Visitor
LHW	Lady Health Worker
MOH	Ministry of Health
MCH	Maternal and Child Health
NGO	Non Governmental Organization
NP-FPPHC	National Program for Family Planning and Primary Health Care
NWFP	North West Frontier Province
PAHO	Pan American Health Organization
PHC	Primary Health Care
PC-1	Planning commission 1
PPIU	Provincial Program Implementation Unit
PC	Provincial Coordinator
WHO	World Health Organization

Chapter: 1

PROLOGUE

1.1. Background

Pakistan is one of developing countries situated in South Asian region with an estimated population of 180 million inhabitants (International database 2009). It is a federation comprising of the federal capital Islamabad, four provinces namely; Khyber Pakhtunkhwa, Punjab, Sindh & Baluchistan and two autonomous regions of Federally Administered Tribal Areas & Gilgit Baltistan. (Map attached in Annexe-1).

Pakistan's health indicators "such as life expectancy of 63 years, infant mortality rate of 85/1000, crude birth rate of 30/1000, crude death rate of 9/1000, high maternal mortality rate, state of prevalence of infectious diseases and poor access to health care facilities are among the worst even in region. The less than 3.8% of allocation for health in national budget compares unfavorably with Iran (7.9%), Sri Lanka (4.8%), Thailand (7.4%), Philippines (4.2%), Bangladesh (4.8%) and Nepal (4.7%). While the government's allocation towards health care has remained well below 1% of the GNP, the total expenditure including that of the private sector has never exceeded 3.5% of the GNP, compared to India with 1.6% of the GNP in public sector and 6% in total. The baseline figure prescribed by the World Health Organization (WHO) is 5% of the GNP" (NP-FPPHC PC-1 2008).

Since independence public health financing in the country has given priority to the curative sector. Although a considerable number of health facilities have been constructed, the rate of their utilization is limited. Recognizing this following the Cairo ICPD Conference, the Government of Pakistan launched in 1994 the "Prime Ministers Programme for Family Planning and Primary Health Care" (PMP-FPPHC) a country wide Programme, under the control of Ministry of Health (MOH) Pakistan, for delivering essential primary health care services to the community through female community health workers. The name of the Programme was later changed to the ***"National Programme for Family Planning and Primary Health Care" (NP-FPPHC)***, but the essential characteristics of the Programme have so far remained the same.

The gender norms related to Pakistan and especially in the north - western Pakistan (including province of Khyber Pakhtunkhwa) restrict women's access to health services. The Lady Health Workers (LHWs) working with this programme are able to deliver appropriate health care to remotest areas. Yet the strict gender norms make the retention of these workers a challenge. The programme aims at improvement of community access to family planning and primary health care services through the LHWs inducted in this programme. These workers provide services at gross root level by visiting the households registered with them, who are a part of their own community. It is expected that this will in turn lead to improvement of all health indicators such as decrease in Infant Mortality Rate (IMR), Maternal Mortality Rate (MMR) and an increase in indices like Contraceptive Prevalence Rate (CPR), Antenatal Care, improved coverage of expanded programme for Immunization (EPI), ORS utilization, health education and utilization of health facilities. This, in turn is deemed to improve impact indicators like as infant and maternal mortality & nutrition of children. Improving the

Contraceptive Prevalence Rate is the way to bring down rate of population growth.

The Government of Pakistan (GOP) recognizes the fact that poverty cannot be eliminated unless the causes of poverty are addressed and dealt with. This change is supposed to result in restoration of economic growth & improvement in access to basic citizen needs (such as primary education, preventive health care & population welfare services). All these measures are essential for winning the fight against poverty (Interim PRSP Pakistan 2002).

This programme has been helping in increasing the overall awareness of the community in order to improve of their quality of life; by measures like having smaller families, self-development initiatives and community social programmes. Further improvement may occur from an inter-sectoral collaboration that can result from the Programme implementation. By improving health status, investments in the Programme contribute to poverty reduction (PRSP 2002). Providing means of respectable livelihood to women in the rural communities of the country in return contribute to the cause of women empowerment. The LHWs also play a meaningful role in promotion of female education.

1.2 Lady health worker of NP-FPPHC Pakistan:

The selection criteria of lady health workers is a result of coordinated efforts of programme authorities, along with local government & involvement of local basic health unit (BHU). This is to ensure opinion of local community in the process. The job of the female community worker includes all the elements of primary health care ranging from healthcare education, Maternal & Child health & provision of basic medicines. In addition to this, they have to carry out other activities like routine vaccination (however this results in an increased workload) (OPM 2002 & PHC wing MOH Pakistan 2006). For the purpose of joining the programme each applicant has to pass through a standardized procedure, and has to fulfill certain criteria which are as follows:

- The female applying for the job should be a local resident of the area where she is going to work
- The age should be 18 years or more while applying for the job
- Preferably married
- Minimum education till Grade VIII
- Ready to work in the field
- Available to complete 15 months of training.

Unlike rest of South Asian countries the training period for community health workers in Pakistan is 15 months, which in turn makes them more capable of dealing with health issues pertaining to the community. Once the LHW is recruited she has to perform certain duties/activities in the field as well as while staying at her home. These routine duties of a Lady Health Worker can be described as follows:

- To provide basic health services to her own community which includes total population of 1000 - 1200 people approximately or 150 to 200 households.
- While providing community health services the lady health worker sets her home as a, "Health House" where charts for health education are displayed.
- She keeps a strong liaison with the local basic health unit.
- She visits 5-7 households daily in her community for which she has an approved tour schedule from the start of the month.
- She registers, all her community in family register as well as pregnant women for antenatal care.

- She registers all the eligible couples for family planning and then provides family planning services accordingly, promoting family planning in order to maximize Contraceptive Prevalence Rate.
- She works in routine immunization activities within her own community (although this has been identified as an increased workload). (National programme PC-1 2003-2008).

The Lady health workers are crucial for strengthening the programme itself, still the programme loses a lot of Lady health workers every year due to various reasons, which are either related to programme itself or to the Lady health workers individually, this loss in technical terms is labeled as *"Attrition"*, which can be defined as *"a factor, normally expressed as a percentage, reflecting the degree of losses of personnel or materiel due to various causes within a specified period of time"* (Dictionary US department of Defense 2005).

Another important term leading to high attrition is **"Resignation"** which implies Number of fully trained Lady health workers that choose to leave the National Programme for Family Planning and Primary Health Care for personal or family reasons, in a stated unit of time (NP-FPPHC HR reports) This survey dealt with the same issue and concentrated on finding out the causes of resignations of the trained LHWs leading to attrition.

The National Programme for Family Planning and Primary Health Care (NP-FPPHC) is federally funded, the current Planning Commission 1 (PC-1) was approved from 2003 – 2008, later extension was given till 2010 (National programme PC-1 2003). The Government of Pakistan, considering its importance and achievements, has decided to continue the Programme till 2011, also including the Programme in its 10 year Perspective Development Plan.

In the National Health Policy of 2001, priority attention is accorded to primary and secondary sectors of health to replace the earlier concentration on tertiary care, with the National Programme for Family Planning and Primary Health Care as its centre piece. This programme has a total workforce of 89,000 lady health workers and it is working at gross root level since 1994. The lady health workers of the programme are now an occupational group that is recognized by the community for services that they are able to deliver. The Lady Health Workers Programme (LHWP) is able to provide services to the community in all Provinces and Regions of the country, providing a total coverage of 63% to the whole country. Full coverage is being planned in next few years (LHW MIS NPFPPHC 2009).

The Lady Health Worker Programme (LHWP) has increased its coverage in the past five years due to increased funding. The Programme has retained its competence in recruiting, training and supervising village based women. The professional knowledge and skills of the LHW and her supervisor have increased, which is a result of improved education level, effective training and supervision (OPM 2009).

Establishment and success of the Programme over the past fifteen years demonstrates that the Programme is valued by community. A high attrition rate however creates several problems like frequent turnover of trained human resource, lack of continuity in the relationships established among lady health workers, community & health system. Huge financial losses are incurred as considerable investment is made on each lady health worker, costing the programme for identifying, screening, selecting, and training the lady health worker. Hence a rise in attrition rate causes wastage of resources.

Lady health workers programme is having an important impact on the health of the population that they serve, however the high attrition rate puts its success in danger. Nationally improvements have been brought in a number of important aspects of service delivery and programme management, but still some important challenges need special attention. When lady health workers leave their posts, the opportunity to build on experience and further develop skills over time through refresher training is lost. Moreover, resignations of these lady health workers leave their community without Primary Health Care (PHC) services for a considerable period of time. Hence frequent turnover rates of lady health workers make the programme less effective.

Through studies it has been demonstrated that although the volunteer programmes are cheaper in terms of salaries, but very high attrition rates still affect these programmes resulting in a need of frequent training of new volunteers. It is difficult to keep track of volunteers and to judge their usefulness (Gilson et al. 1989). Retention is affected by central concerns with governance and management; such as sources of financing, community ownership and selection practices. For this reason retention of healthcare workers can and should be addressed as part of a broader package of management interventions.

Chapter: 2

COMMUNITY HEALTHCARE PROGRAMMES AROUND THE WORLD

2.1 General information about the Human resource situation in different countries

Virtually all developing countries in the world are facing challenges related to their human resources situation. The shortage of health workers in these countries is due to a number of reasons. The most common reason being a faster population growth rate with a lesser ratio of healthcare workers to serve growing population. Other main reason is lack of resources in developing countries due to which the health workers are not paid properly, resulting in decreased motivation and eventually attrition. Another important reason for shortage of health workers is their decision to settle and find jobs in more settled areas compared to periphery & hard areas. This creates an imbalance in distribution, resulting in shortage of workers in already deprived areas. These problems related to workers shortage exert a negative impact on quality of health care, control of diseases and eventually achievement of Millennium development goals (PAHO, WHO 2007).

Estimates from studies show that there are 59.2 million fulltime paid health workers worldwide (WHO 2006), among which "21.7 million reside within the Americas. Still there are an estimated 57 countries worldwide with critical shortages in healthcare human resource" (WHO report 2006). A recent review of healthcare human resource trends in the Americas, estimates the numbers of health services providers in the region to be 12.5 million. Some of the crucial issues related to retention of health providers are:

- Loss due to migration, as noticed to be over 72% in different countries of Americas.
- Higher attrition rates of more than 75% are seen in doctors, nurses and other health professionals in 15 countries where the healthcare human resource density ratio is below 25.
- The ratio of unemployment in women was more in most of the countries.
- In a sample of 13 countries throughout the Region, the average unemployment rate for health workers was 6.2%, with a high of 16.8%" (PAHO 2006).

The results of various reviews repeatedly demonstrate that presence of adequate number of workers in health care has a positive impact on health status of the population, especially in the field of maternal and child health (Lancet 2004).

Studies carried out in countries of Sub-Saharan Africa also suggest that issue of healthcare workers shortage needs more attention in countries with poor health indicators (WHO report 2006; Narasimhan.V et al 2004). According to WHO, to achieve the targets of millennium declaration, the number of health workers in most affected countries of Sub-Saharan Africa needs to be increased (UN MDG 2009). In these countries there is difficulty in recruiting large number of health workers. Additional funds provided for this purpose are not utilized efficiently. In Sub-Saharan Africa, the commonly reported causes of health workers shortage include migration of the health workers, shortcomings in the initial trainings, retirement before the actual time, ill heath and death. (Zum P & et al 2005;Mullan F 2005).

Taking into account the inflows and outflow ratio "the estimated net growth rates of the health workforce to population growth rates can be estimated for respective countries. The current trends should indicate whether the net growth rate of the health workforce is faster than that of the population, allowing health workers density to increase over time. Secondly the extent to which the workforce would need to grow in each country to attain the maximum density of health workers per 1000 population" (WHO report 2006).

2.2 General information on Volunteer/Community based programmes.

The concept of volunteer programmes started developing in mid 1970s, the main reason being recruitment of willing workers while not burdening the national exchequer. The volunteer workers usually are not paid a fixed monthly salary however they retain a small profit from sale of drugs prescribed for common illnesses, or are given some material incentive in form of machines or animals to earn their livelihood.

In most volunteer programmes no material incentives for either health staff or volunteers are there. This on one side, make the cost of the programme extremely low, but on the other hand, sustainability of the programme rests on contact with health staff, seminars, health exhibitions etc. (Walt et al 1989). Although volunteerism is by definition a free service, but for the purpose of earning livelihood regular incentives are considered to be must. For this purpose, every country running a volunteer programme has a regular process of giving either monetary or nonmonetary rewards to their workers. These rewards help the volunteer worker in getting permission from their families for the job, improve their status in the community and are also a compensation of time which they spend on duty (UNICEF 2004).

Evidence has demonstrated that in case of voluntary health services, nonmonetary rewards are better and reliable options compared to payments because while dealing with funds there can be delays in disbursement, which can in turn be a result of delay in provision of funds. Moreover there is always a chance that the funds will stop suddenly, as was the case in India and Nepal, especially if the healthcare workers are in a volunteer based service programme. It is also reported in studies that the Community Health Workers are more satisfied while providing curative services compared to preventive services. To retain these workers in jobs, proper training with increased community acceptability and transparent selection process are required (UNICEF 2004).

The community health worker is known by different names in various countries. In order to strengthen their scope of work most of the community workers have a village health committee (VHC). The VHC also play an ambiguous role within community health worker initiatives. However as "Due to hierarchies within the village, the position of VHC is not always clear and is sometimes contested, leading to build up of tension among VHC and community leaders, resulting in political contestation" (Sanders, 1992; Sauerborn, Nougtara & Diesfeld, 1989; Streefland, 1990; Ebrahim, 1988). Due to cultural reasons, in lady health worker programme of Pakistan, such village committee is divided into two groups, namely male health committee and women group. Nomination of the members of health committee and women group is on will of concerned lady health worker. (NP-FPPHC PC-1 2008).

Supportive supervision, an efficient logistic system along with good infrastructure & strong health system support are essential for Community health worker's effectiveness. In addition, regular supply of essential medicines along with provision of transport, are required to facilitate the Community health worker's work. "As the Community health workers mostly operate in periphery, both geographically & organizationally, they are the first to lose training opportunities and supervisory visits" (Gilson et al., 1989).

In case supplies of drugs are delayed, the health workers cannot perform their job properly. This adversely affects their acceptability and respect in community. (Ofosu- Amaah, 1983). "If CHWs are working in programmes that have drug treatment as their core objective, such as TB DOTS or HAART, the situation becomes even more critical" (Farmer et al., 2001), most programmes include the need for supply of drugs and/or equipment, including transport. (SOCHARA, 2005).

Most of health personnel lack the awareness of providing a supportive environment for Community health worker programmes. Much of their work orientation is towards clinical side and thus, have a poor concept of primary health care. It is for the reasons; they are unable to provide a supportive environment of partnerships and teamwork. Community health workers are sometimes considered as low grade employees by them (WHO, 1989; WHO, 1990; Walt, 1992), their health promoting and enabling role is misunderstood by the health professionals, so they want them to work as assistants in their respective health facilities. A sense of superiority of health personnel has been observed as a problem (Sanders, 1992), however this problem needs to be addressed at basic levels like training of medical students (Waterston & Sanders, 1987).

The role of village health worker in the past was not only providing health care, but they were also considered to be an advocate for community, act as agents for social change, and work against inequities & for upholding rights of community (Werner, 1981). The most important "role of a Community health worker is to act as a bridge between the community and formal health services in all aspects of health development". According to the definition of The WHO Study

Group: *"Community health workers should be members of the communities where they work, should be selected by the communities, should be answerable to the communities for their activities, should be supported by the health system but not necessarily a part of its organization, and have shorter training than professional workers"* (WHO 1989). Another definition used in literature is: *"any health worker carrying out functions related to health care delivery; trained in some way in the context of the intervention; and having no formal professional or paraprofessional certificated or degreed tertiary education"* (Lewin et al). The training of Community health worker is a recognized by health services and national certification authority, but is not a part of tertiary education certificate" (WHO 2007)

There have been several definitions of the term "Community Health Worker" within a large number of different titles (Bhattacharyya & etal), different variety of names are used for these workers such as "Health Auxiliaries, Barefoot Doctors, Health Agents, Health Promoters, Family Welfare Educators, Health Volunteers, Village Health Workers, Community Health Aides, Community Health Volunteers and Community Health Workers. With the varying demands and differing levels of health within countries, regions, districts, and villages, each community has its own version of the community health worker" (UNICEF 2004). Some of the titles used for these community health workers in this study are given in Table 1. (Bhattacharyya et al, 2001; Gilroy & Winch 2006).

Country	Name for community worker
Pakistan	Lady health worker
Nepal	Female community health volunteer
India	Basic health worker, Anganwadi
Bangladesh	Shastho shebika, Shastho Karmis
Various countries	Health promoter, outreach educator, Village health

Table 1: Names given to Community Health Workers

2.3 The problem of attrition in Volunteer programmes.

Volunteer workers "were originally conceived as ones having a broad role as agents of development, spreading community participation within their own communities, as well as being educators and communicators". Although the issue of attrition is common to all community based programmes, but is more frequently noticed in volunteer programmes, the main reason for this being absence of monetary incentives which lead to lack of interest on part of the worker, ending up in attrition.

The term Attrition has already been mentioned in Chapter: 1 as "a factor, normally expressed as a percentage, reflecting the degree of losses of personnel or materiel due to various causes within a specified period of time". It has some subdivisions in the context of Lady Health Worker program such as;

i. **Resignation:** All fully trained Lady Health Workers who choose to leave the program due to personal or family reasons, in a stated unit of time.

ii. **Drop-out:** All the Lady Health Workers who left the programme due to relocation or resignation in a stated unit of time.

iii. **Relocation:** Fully trained Lady Health Workers that move away from their area to live outside of their original community so they are no longer permitted to continue their work, in a stated unit of time. This is mostly due to marriage.

iv. **Termination:** Lady Health Workers who are asked to stop work due to poor performance, in a stated unit of time.

v. **Total attrition:** Lady Health Workers who left the programme in a stated unit of time.

(All the definitions mentioned above are from NP-FPPHC HR records).

2.4 Examples from all over the world

In the following pages, examples of community based initiatives will be presented and information will be given as to what extent, these programmes suffer from high attrition. "The concept of community members to render certain basic health services to the communities is fifty to sixty years old, the *Chinese barefoot doctors* date back to 1950s, at the same time, Thailand also made use of *village health volunteers*" (Kaufmann & Myers 1997; Sringernyuang et al 1995). The success of different community based programmes and the failure of conventional allopathic health services to deliver basic health services induced a number of countries to try the concept of village health workers. (Sanders, 1985).

Most of the South Asian countries piloted community health worker programmes in late 1970s which were further expanded in 1980s. From 1980s onwards, most of the community health worker programmes declined mainly due to global economic recession, political instability, neo-liberal economic policies and difficulties in financing the programmes. Community health workers also received little support in terms of training management & supervision but the volunteers often dropped out due to lack of motivation. Evidence suggests that well organized community based programme can really bring about a change.

In **Bangladesh,** NGO BRAC is one of the largest organizations having their unsalaried community health workers "(called Shastho Shebikas). The Shastho Shebikas are members of the BRAC village organizations, that consist of women from the poorest communities, and which are aimed at improving their socio-economic conditions. They are not paid a salary but they retain a small profit from the sale of drugs prescribed for common illnesses "(Khan et al 1998, Hossain1999).

Evidence from Bangladesh demonstrates "that the dropout rate for community worker ranged from 31 percent to 44 percent. Some dropped out after only a few months while others did so after a few years of service (Khan et al. 1998). The main reasons for discontinuing the work were as follows:

- The profit from sale of medicines was less, they could earn better in other jobs.
- Were not able to give enough time to the job because of family matters.
- Too much effort spent for too little profit.
- Could not find enough customers to sell medicine in order to target set by office, which was usually too high to be achieved.
- No salary.
- People were not ready to pay cash money but wanted to buy on credit instead.

- Population being reluctant to buy medicine because of their perception that BRAC got medicine for free.
- The medicines in the local market were cheap as compared to the ones available with the community health workers.
- Preference for buying medicines from local shops.
- Family members disliked the work.
- Socially unacceptable for a woman to do this work.

In Bangladesh's BRAC programme, Community health workers quit their jobs due to lack of time, lack of profit, and family's disapproval. The results of dropouts not only affected achievement of targets, but also incurred loss of money per dropout for their training and supervision" (Khan et al 1998). There are multiple reasons for worker's attrition in this programme but the most common problems are related to initial recruitment process of these workers. "Another frequently cited reason for high attrition rate was movement to higher positions in the health system, marriage or family matters, and finding better positions in other fields" (Ofosu-Amaah 1983).

Bhutan is another country which has community based programme. The programme was launched in 1979. The basic philosophy of this programme is to establish a link between the community and the health services. There is no direct input from the government into Village Health Workers programme, "except the supervision and support given to them through the decentralized health system. There is a very high attrition rate and there is no specific mechanism in place to check this problem (UNICEF 2004).

In order to strengthen the role of their community based workers the Bhutanese government made a clear cut policy for selection, payment, training and supervision of Volunteer Health Workers (VHWs). It was observed that five year attrition rate for Volunteer Health Workers in Bhutan was 50 to 55%. Individual districts had attrition rates ranging from 21% to 63%. The most common reasons cited for attrition were:

- That the work interfered with their personal work (70%).
- Family pressure (12%).
- Too hard job (9%) and there was nothing to be gained from it (6%).
- Bhutanese government role being indirect, it would not be very surprising if the majority left as there was no remuneration and career prospect as a Volunteer Health Worker.

The attrition rate of health workers in Bhutan was high the main reason being voluntary nature of job. Still among issues needing clarification include, the policy for recruiting replacements for Volunteer Health Workers who resign (Danida 2001).

A similar programme in **India** started working in 1977. The concept of this programme was same that is *"provision of health services at the doorsteps of villager"* (Chatterjee 1993, Maru 1983). In India, the name of the community worker changed from time to time. In the start of programme it was Community Health Worker later their name changed to Village Health Guides (VHGs). Village Health Guides in India encountered a number of difficulties namely;

- Inadequate support from their communities and the health system.
- Another issue enveloping the VHGs was their 'medicalization'.

After being trained for three months, they focused on providing curative services to the neglect of preventive and promotive tasks. This was due in large part to their orientation to curative care during their initial training, which was conducted by Primary Health Centre doctors and Health Supervisors. These professionals were themselves not instructed appropriately on how to train basic health workers. (UNICEF 2004).

In India the community health workers programme was a failure, reasons being lack of ownership and support by the central government. The salaries of the Village Health Guides were stopped suddenly and the programme was closed after running for 25 years. The failure of Village Health Guides programme in India shows that for sustainability of community based programmes, strong ownership by the government and the community is needed.

The National Female Community Health Volunteer (FCHV) programme in **Nepal** was introduced in 1988 by the Ministry of Health and Population. Female Community Health Volunteers cover all communities in the country, and generally are responsible for 100-150 households. They do not receive pay for their community service however their contribution is recognized in various ways by government and their communities. Attrition rate of these workers was found to be very low i.e. less than 4% per year (UNICEF 2004).

The total number of Female Community Health Volunteers is almost 49,000 throughout Nepal. Most are not literate, but they serve as a gross root health resource, providing community based health education and services in rural areas. They distribute vitamin A capsules biannually, work in National Immunization Days, distribution of family planning products, and oral rehydration salts. They also provide community based treatment of acute respiratory infections (including use of antibiotics) and referral to health facilities in programme districts (USAID report 2008).

In Nepal too, there are problems with retention of community workers, but in order to reduce attrition, the government through orientations and review meetings, made efforts to generate support from local government bodies and communities in organizing mother's group meetings, replenishing Female Community Health Volunteer's medicine supplies, disseminating information and providing incentives.

For the retention of workers, in addition to remuneration and incentives, community recognition and public appreciation for their contribution as volunteers in form of awards, certificates, ceremonies, etc. are desired. These measures were identified by volunteers as important factor contributing to their own sense of satisfaction & motivation to continue (Government of Nepal and Maternal and Neonatal Health 2003). Nepal has a good example of community involvement and ownership. They have developed Village Development Committee which collects funds for further paying it to their Female Community Health Volunteer.

Sri Lanka is another example where involvement of school teachers and village/community leaders in voluntary community work dates back to 1915, another example is that of *"Rockfeller Foundation sponsored campaign for control of hookworm infestation"*. During the malaria epidemic of 1934/35, voluntary workers participated in the control activities. The non-government organizations on the other side began involving volunteers from the 1950s onwards (Walt 1989).

In **Solomon Islands,** Attrition is more common with young age and irregular payments, 66% of the non working Volunteer Health Workers surveyed had quit because of pay-related reasons. These young workers abandon job of Volunteer Health Worker when they get married and find another employment to support their families. Some of the Volunteer Health Workers considered their job as a steppingstone to becoming a Nurse. The local communities tried to have their Volunteer Health Worker posts upgraded to that of nurse's aide. If this practice was restricted, more Volunteer Health Workers would drop out. Attrition of community workers in Solomon Islands was attributed to causes like inadequate pay, family reasons, lack of community support, and upgrading of health posts" (Chevalier et al., 1993).

Higher attrition rates for Community health workers of 3.2 percent to 77 percent are reported in literature, associated mostly with volunteers as discussed above in context of South Asian countries. Attrition rates of 30 percent over nine months in Senegal and 50 percent over two years in Nigeria were reported (Parlato & Favin, 1982). The Community health workers who receive government salary were more often prone to retain their jobs as compared to ones who depended on community financing.

In **Sub-Saharan Africa (Nigeria)**, where onchocerciasis is endemic, ivermectin was distributed annually through community directed distributors (CDD), as part of the effort to control the disease. It has been reported that at least 35% of the distributors who have been trained in Nigeria are unwilling to participate further as CDD. The selection and training of new Community workers to replace the unwilling workers, lead to annual expenses that the national onchocerciasis programme is finding difficult to meet, because of the programme having other priorities and limited resources. If the reported levels of attrition are true, they seriously threaten the sustainability of community directed treatment with ivermectin (CDTI) in Nigeria.

In 2002, interviews were held with 101 people who had been trained as CDD, including those who had stopped serving their communities, from 12 communities in south-eastern Nigeria that had high rates of CDD attrition. The results showed that, although the overall reported CDD attrition was 40.6%, the actual rate was only 10.9%. The CDD who had ceased participating in the annual rounds of ivermectin reported the following problems:

- Lack of incentives (65.9%).
- Demands of other employment (14.6%).
- Long distances involved in the house-to-house distribution (12.2%).
- Marital duties (7.3%).

"Analysis of the data obtained from all interviewed CDDs showed that inadequate supplies of ivermectin, lack of supervision, and a lack of monetary incentives lead to significant increases in attrition" (Annals of Tropical Medicine January 2008). Each country mentioned in the literature has learned its own lessons and witnessed significant changes in focus during the course of programme.

2.5 The Lady Health Worker Programme in Pakistan

Like the rest of the South Asia, the government of Pakistan also launched the "National programme for family planning and primary health care" famous by the name of "Lady Health Workers programme" in 1994 for improving maternal and child health and overall health indicators of the community, both in rural areas as well as urban slums. The programme has a huge force of female community workers and their female supervisors who supervise them in the field (MOH GOP 2007).

Lack of awareness and poverty amongst the people is a critical issue, this problem is taken up by the Lady Health Workers (LHW) who visit their community door to door both in the villages and the urban slums, arrange awareness campaigns and provide essential medicines for treatment of minor ailments amongst the poor. The LHW mobilizes the community to promote and improve health through her participation in the village health committee (male) and in the women's health committee. In Pakistan and other low income countries, the provision of primary health services is designed in a way that provision of health services is ensured mostly at gross root level.

The National programme for family planning and primary health care is one of the largest community based programmes in Pakistan, and is owned by the government. It provides primary health care services to 63% population of country (NP-FPPHC MIS 2009). The programme is working successfully and shares many common features with similar programmes in other South Asian countries. Over the past few years, the Lady Health Worker Programme (NP-FPPHC) has become an important element in the Government of Pakistan's plan to improve the health status of women and children in rural areas and urban slums.

The evaluations show that these services are exerting a positive impact on health of the poor, particularly women and children. The LHWs are contributing directly to improved hygiene and higher levels of contraceptive use, iron supplementation, growth monitoring and vaccination amongst their clients. Almost three out of four communities report that the LHW has generally improved people's lives (OPM report 2003). LHWs also see patients with emergencies referring them to appropriate health facilities in addition to providing preventive and promotive services. Unfortunately clinical referral support services at the health facilities throughout the country are inadequate which limits effectiveness of the LHW's referral role (OPM report 2003).

Since the commencement of LHW programme in Pakistan, a

number of programme evaluations have been carried out both internal and external. Right now the feedback report of fourth evaluation has been given to the programme but not actually published, till now the most extensive among these was the national level, external evaluation conducted during 2000– 2001 (OPM 2003). It concluded that the programme succeeded in creating a large sized organization comprising female community health workers and establishing a functional programme management & supply system (OPM 2009). It was found that the programme improved the uptake of important health services in areas covered by its LHWs.

It has been reported in the fourth evaluation report that majority of LHWs have access to a mobile phone, which contributes to better service provision. Across the country, seventy six percent of LHWs now have access to a mobile phone. Many of the LHWs have shared access with their husband or another family member. In Khyber Pakhtunkhwa nearly *eighty-two percent* of LHWs make use of a mobile phone for giving timely services to their community (OPM 2009).

Multiple studies have been conducted on different aspects of the LHWs programme, but none has so far to highlighted LHWs own views about their job description and occupational stress. These factors would be important in bringing about improvement in quality of service delivery and overcoming under performance/utilization of existing LHWs. This issue in general remains poorly researched but some reviews done so far have pointed out knowledge gaps in areas like job satisfaction/dissatisfaction and job retention/attrition (WHO report 2005; UNICEF SA 2004).

Celebrating its success, the President of Pakistan announced 2008 as the "Year of Lady Health Workers", with the intention of increasing their respect in the country, and also to decrease attrition of the programme (NP for FP & PHC 2008).

According to reports of third external evaluation of National programme for Family Planning and Primary Health Care Pakistan, has an attrition rate of 5% (OPM 2003). These attrition figures were taken into account from the total number of working LHWs in the whole country at the time of evaluation and all causes like termination, resignation, dropouts together were included in it. This means additional 5000 LHWs are needed to be trained annually, costing another loss of 4 million Rupees to Government of Pakistan. (OPM report 2003; MOH GOP PC-I 2003-2008).

The Government strategic plan for 2003 – 2008 aims to build capacity at lower government levels, by trying for more sustainable structures of management, in order to avoid the workers losses (MOH 2003 – 2008). It also has a set recruitment targets of 110,000 LHWs till 2010, to cover 30% of the urban and 90% of the rural population (PHC report 2005-2006). The main reasons for this being community demands in the areas which are not covered by the LHWs programme.

It is challenging to sustain a huge programme like this in a country like Pakistan, where gender norms are strict, making going out for job, entering other households and discussing health issues like reproductive health a challenge for these female workers. On the other hand the programme only inducts female workers, which is one of its big strengths, as it makes the community household visits easier. Also interaction within the community and organizing health awareness campaigns becomes much easier. It is seen that articles on Community health worker programmes in Pakistan and Bangladesh specifically mention the gender of their health workers, compared to articles on such programmes in Latin America and Africa, gender not being a major cultural issue in those countries. This clearly indicates that gender issue to very large extent is influenced by wider societal practices and beliefs. (WHO report January 2007).

2.6 Other reasons for attrition

With the passage of time there is reduction in support by development partners, to community based programmes, as these programmes are perceived as reasons being taken as "just another pair of hands" in provision of health services. For this reason strong support from community and government is needed to sustain their network (Sharma 2003). Evaluation of programme had identified a number of problems related to selection criteria of Community health workers (UNICEF 1993/94). The studies also comment on; time needed by these programmes to perform, their supervision, limited opportunities for on job training, no scope for career advancement, poor compensation and unpredictable sustainability" (Danida 2001).

In some countries like **Peru**, majority of health promoters & indeed traditional birth attendants are male. This also appears problematic because they dramatically skew gender equality in community leadership positions. Resistance from husbands has been identified as a key barrier to the participation of women in these programmes (Brown et al 2006).

Studies also suggest that some incentives for addressing worker productivity and retention may be more favorable to female than to male workers, such as flexible working hours and leave arrangements (Gupta & et al 2009). In addition to working conditions, factors affecting women workers should be more critically analyzed; such as physical workloads, reconciling work and family, relations with clients and sexual harassment.

Sexual harassment is a major contributing factor of attrition in community workers. Sexual harassment implies; a request for sexual favors, sexual advances which are not welcomed, and other verbal or physical conduct which hinders work and creates a hostile or offensive working environment (Rubin PN 1995). The most common form of harassment experienced by the workers is compliments on body and/or figure with flirtation & sexiest jokes. Unwanted sexual contact and explicit sexual proposition are also reported in health care. In case of community workers, they have to face harassment in community, or at the hands of male trainers, or supervisors who threaten them of terminating their jobs (NP-FPPHC 2008-09). In male dominated societies the prevalence of sexual harassment influences the female workers employment retention rates.

Psychological violence, physical violence & verbal abuse are also very common in workplace and may lead to attrition. Evaluation of environment that creates risks for violence is necessary to guide the formulation of meaningful interventions (Jackson M & et al 2005).

Another major cause of attrition of health workers in South Asia is continuously deteriorating security situation. Same is true for Afghanistan and Pakistan. Recently in Pakistan, security threats are faced by health workers and in some cases, factual killing of the health workers have also been witnessed. An official working for the United Nations High Commissioner for Refugees (UNHCR) and his bodyguard were shot dead by some gunmen during a failed abduction attempt in the northwestern Pakistani city of Peshawar (World news 2009). Regarding the NPFPPHC safety problem are more oftenly faced by unmarried workers compared to married ones. Male guards were also requested by some of the LHWs for pick and drop services from Basic Health Units as it was not deemed safe to travel the distance alone (Harriet burn 2008).

Based on the literature reviewed, it was noticed that each country has its own view for the community health workers programme setup and functioning. But there are few things in common, related to roles and responsibilities:

- "Increased responsibility lies on local governments.
- Involvement of community is necessary in treating basic health problems.

Except India, all countries reviewed, are still continuing with Community health worker programmes and have integrated them into national health services programmes (UNICEF 2004). For improving the services related to primary health care, it is necessary that each country have a well established community health workers programme, which should be owned by their respective government with a regular mechanism of disbursement of salaries in place to improve worker's satisfaction and hence retention.

Chapter: 3

THE SURVEY

Before conducting this survey, some background formalities were carried out, First of all a formal request for conducting a survey was sent to programme authorities in order to get permission. This was done to avoid any official issues later on. Following this, informal meetings with Provincial Coordinator (PC) & Assistant Provincial Coordinator (APC) of the programme were arranged, to create a sound knowledge of official procedures required and to design all the activities in a structured way. After the meetings the APC issued an official letter of permission to the concerned districts, asking for required support.

3.1 Design

This survey was a simple descriptive study which is qualitative in nature. This design was selected because it can cover the objectives within allocated time period and is cost effective. This survey was done with an intention to investigate causes of resignations of fully trained Lady Health workers (LHWs) who resign on their own will, which is an important factor of attrition in National Programme for FP and PHC. This survey was also attempted to assess LHW's knowledge regarding their job, how they perceive themselves in their communities. Along with this associated factors which influence attrition and retention were also looked for.

3.2 Data collection methods

For the purpose of conducting this survey, two separate

questionnaires were developed by the principal investigator and following data collection tools and methods were used to achieve the objective of the study

- Document review
- Interviews with the LHWs
- Interviews with the officers

Initially, focus group discussions (FGDs) were also in mind as part of methodology, but due to poor security situation it was not possible to gather all LHWs who had resigned at one place. For Lady Health Supervisors (LHSs) due their busy schedule and additional workload with regards to Internally Displaced People (IDPs), the district managers were not ready to spare them for FGDs, In order to assess opinions of LHSs, they were included in the interviews for officers.

3.2.1 Document review

Document review was also done as an essential part of methodology, as it is impossible to reach to any conclusion without review of the relevant data. For this purpose a meeting with the District coordinators (DC) and Assistant District Coordinator (ADC) was arranged; to explain the purpose of survey, with a request to provide the necessary records/data of all the Ex-LHWs. The idea behind conducting this review was twofold, firstly, to find out proportion of LHWs who have resigned willingly from total causes of attrition, and secondly to find out any possible administrative or official cause of LHW's resignation. For this reason a document review tool was developed in order to prevent missing any necessary information (copy attached in annexure 8.5).

Documents reviewed included relevant data, like district monthly reports, annual reports, payroll and personal profile of the LHWs. The documents were reviewed in detail and following inormation about the LHWs was collected;

- How many among all cases of attrition have resigned by their own will?
- Was there any administrative action in the form of explanation calls/warning letters /show cause letters available in the personal profile of the resigned LHWs?
- If yes, how many have resigned soon after administrative action was taken against them?

All LHWs who had resigned were traced by records available at the Provincial Programme Implementation Unit (PPIU) Peshawar Administration Section, and the District Programme Implementation Unit (DPIUs) of the districts. The following information was obtained from document review:

Mardan district:

Total attrition of LHWs (all causes)	54
Total resignations on will	26

Nowshera district:

Total attrition of LHWs (all causes)	45
Total resignations on will	11

It was also found from document review that in total 39 LHWs from both districts had resigned from their job on their will (however 2 out of them were not interviewed and the rest agreed to it). (Districts Profile attached in Annex- 8.2)

3.2.2 Interviews with LHWs

As mentioned earlier basic information about the resigned LHWs attrition was obtained from DPIUs of the study districts. However some of it was counterchecked and confirmed for accuracy in the Provincial Programme Implementation Unit (PPIU) Peshawar, and final confirmation was done through PPIU finance section records by checking the payroll.

A semi-structured questionnaire was designed for interview with LHWs however the questionnaire was pretested in the district Peshawar which was not the study district. The main purpose of pre-testing was to check sequence of questions, timing of interview, coding and to see respondents understanding. Problems identified were corrected. Later after making corrections, the revised questionnaire was again pretested at the same place. A manual of instruction for efficient questionnaire administration was designed before the pre-testing and later updated in the light of pre-testing. It was a guide in the data collection. The initial approved questionnaire was in English, later for the local adaptation it was translated in Urdu.

As all the 37 LHWs who had resigned were living in different parts of the study districts, it was difficult to visit these areas alone, for this purpose five LHSs were hired for a period of three weeks and trained on taking the interviews. For this purpose a one day training session on the questionnaire for LHWs was conducted by the principal investigator in both the study districts. During the training session all the questions were discussed in detail, along with the technique of asking supplementary questions and interpretation of the answers. The importance of assurance of confidentiality was also explained, because these LHWs were no more programme employees. Moreover confidentiality was one of the norms of this study. (Questionnaire attached in annexure 8.3)

For the purpose of interview the resigned LHWs were first contacted telephonically or either in person, informed consent for the interview was obtained, and the schedule of interview was also finalized. Following the finalization of interview schedule, LHWs houses were located which was either done with the help of LHSs or the DPIU drivers. Finally Interviews were conducted from those LHWs who gave written consent for it (All the LHWs who were interviewed gave the consent without hesitation).

3.2.3 Interviews with officers

The other arm of this survey was face to face interviews with officers of the programme, who dealt with management and human resource issues. An approved semi-structured questionnaire was used for this purpose. For the interview only those officers were selected who had at least 3 – 5 years of experience working with the programme & were willing, available and ready to answer the questionnaire (Sample of the questionnaire attached in annexure 8.4). All officials interviewed were concerned with implementation of the programme and had enough experience working in field. For this purpose the Provincial Coordinator (PC), Assistant Provincial Coordinator (APC), District Coordinators (DCs) of Mardan and Nowshera, and one Lady Health Supervisor (LHS) from each district were interviewed.

3.3 Study Population and Sampling

This survey was conducted in province of Khyber Pakhtunkhwa, Pakistan exploring one of the major public health issues i.e., causes of attrition of the LHWs of NP-FPPHC. The province consists of a total 24 districts, with a total number of 12,500 working LHWs. Mardan and

Nowshera are two main districts this Province. As there is a deteriorating security situation all over the province, only those districts were selected where law and order situation was comparatively acceptable and travelling to the periphery was possible.

At first, District Peshawar which is the capital city of Khyber Pakhtunkhwa was selected for this purpose, however due to security reasons and lack of available information; it was replaced by the districts of Mardan and Nowshera. The programme is well established in these districts, operating on a large scale and the areas being logistically manageable. The total population coverage by the LHWs, in Nowshera district is (71%), and in Mardan district is (68%) (See table)

District	Rural	Urban	Total working
Mardan	1089	136	1225
Nowshera	755	20	775

Table: 2 No. of LHWs districts Mardan & Nowshera in 2008

After gathering complete information on all causes of attrition like resignation, termination, dropout etc, only those LHWs were purposefully selected for the study who had resigned willingly in year 2008.

3.4 Data analysis

The data was gathered by Principal investigator with help of five field workers. The first field editing was done by interviewers whereas second field editing was done by the principal investigator.

Data was interpreted by reflecting on findings, recognition of commonalities and identification of any pertinent links or associations. Well-trained data entry operator was hired and data was entered in Statistical Package for Social Sciences (SPSS Version 17) programme. Data was validated by double entry.

3.5 Limitations

The study was planned in province of Khyber Pakhtunkhwa which is a victim of deteriorating security situation hence conducting it in field setting was itself a challenge. The main problems faced during the course can be summarized as follows:

3.5.1 Security reasons

At first the district for survey was proposed to be city of Peshawar, which is the capital city of Khyber Pakhtunkhwa, however the security situation became worst at the time of, so the area was changed to districts of Mardan and Nowshera, which were comparatively safe, But till very end there was no surety that the work can be conducted safely

3.5.2 Non availability of relevant data

Another reason of shifting of survey from District Peshawar was that; the records were not being maintained properly. Some necessary information was missing, without which it was impossible to reach the LHWs.

3.5.3 Absence of Interviewer

Due to security reasons it was not possible to train anybody from outside the programme for interviewing or accompanying the principal investigator. The only option left was LHSs of the programme who were trained for this purpose, and were asked to visit houses of Ex-LHWs with the principal investigator. These trained LHSs sometimes refused to come on the scheduled days due to commitments like personal problems or busy schedule (But in reality they had reservations on going to remote areas because of ongoing security situation). This sometimes resulted in disruption of interview schedule.

3.5.4 Long distances

As there were two districts in which interviews were scheduled, and the interviews were conducted in houses of the LHWs who had resigned from programme, the investigators had to travel long distances to reach them, sometimes this was unsafe and time consuming.

3.5.5 Personal Bias/ field Bias

By training the LHSs of the programme on LHWs questionnaire there was a chance of creating bias in the results, because some of the questions in the LHWs questionnaire were related to the LHS visits and behavior with the LHWs where they could have made wrong interpretation of the actual answers. To avoid any bias in the results the principal investigator was present in all the interviews.

3.6 Ethical consideration

For the smooth conduction of the survey and to avoid ethical issues a formal letter was sent well before the time of actual conduction of survey, to federal and provincial programme implementation units. This was done knowing that it is not liked by programme managers, to go directly to the field offices and use the programme data for personal purposes. Moreover, with any research it is important to keep the well being of the participant a top priority, and it should not harm the participants at any cost. The survey was planned sensitively keeping in mind the relevant problems; confidentiality, anonymity and privacy were constantly maintained by both the researcher and others members involved. Only females were involved for interviewing keeping in mind the cultural norms of the area. All the LHWs were approached at their houses and not anywhere else.

The informed consent document was translated in local language so that it can be easily read and understood by the respondents. The contact details were provided to all the participants in case anybody wanted a follow up. Consent for the audio recording of interviews was requested, but all participants refused to allow so because of social and cultural reasons. Moreover they were afraid of their voice being recorded and later being misused. (Ethical note attached in Annex 8.6)

Chapter: 4

INFERENCES

The findings of this survey have been organized in the following categories; First part consists of results of information obtained by interviews of LHWs who had resigned from their positions. The second part demonstrates findings of the interviews with officers of LHW programme.

A total of 37 interviews with the LHWs who had resigned were conducted in two separate districts. Table: 3 demonstrate basic demographics of lady health workers who had resigned willingly from the programme.

The mean age of respondents was 23.66 years. The lowest age was 18 years and highest age was 43 years. At the time of joining 73% of LHWs were unmarried and at time of interview 43.2% were unmarried, which implies that marriage is an important cause for resignation of LHWs. This is a major concern for programme authorities which should be addressed by taking some policy initiatives. Among the LHWs no one was divorced or separated, except one widow. Majority (51.4%) of the LHWs had passed high school (Grade X) while 18.9% had attended college (Grade XII), this fact demonstrates that the literacy level of the LHWs was good, because most of them were lying in the category of high school as compared to secondary school, but again better literacy rate can be a cause of attrition as they are more likely to get better jobs (e.g. in schools as teachers) (Table: 3). Major responsibilities of LHWs are highlighted in Table: 4.

S.No	Variable	Description
1	Total no. of LHWs interviewed District Mardan District Nowshera	37 26 (70.2%) 11 (29.8%)
2	Age of LHWs 18- ≤30 years 30-45 years Mean Age	 32 (91.4%) 03 (08.6%) 23.66 years
3	Educational Status Secondary school (Grade VIII) High school (Grade X) College (Grade XIX)	 11 (29.7%) 19 (51.4%) 07 (18.9%)
4	Marital Status <u>At time of Joining</u> Unmarried Married Widow	 27 (73.0%) 09 (24.3%) 01 (02.7%)
5	Marital Status <u>At time of interview</u> Unmarried Married Widow	More 16 (43.2%) 20 (54.1%) 01 (02.7%)

Table 3: Basic Demographic of Lady Health Workers

S.No	Variable	Description
1.	Health education	98%
2.	Health of Mother and Child	100%
3.	Registration of Household/Families	98%
4.	Awareness on Vaccination	97%
5.	Awareness on Family Planning	96%

Table 4: Knowledge of LHWs about their Responsibilities

Table: 4 demonstrate whether the LHWs are fully aware of their role and responsibilities of their job. The results in this aspect were very encouraging demonstrating that majority of the LHWs were fully clear about their job related duties. All the respondents mentioned health of mothers and children as their basic responsibility and did mention details in their interview. A total of 98% LHWs mentioned health education and registration of household/families as their responsibility. Awareness on vaccination and family planning was mentioned by (97%) and (96%) of the respondents.

With regards to self perception of LHWs about the opinion of community regarding the duties and role of LHWs, Figure: 1 demonstrates the results. Large majority (89.1%) considered themselves as fully respected by the community which was expressed by the statements like;

i. people gave respect

ii. people were happy because of our work

iii. people were happy because we delivered health services to them

Two respondents (5.4%) had initially experienced difficult beginning of their relationship with the community, but later it changed to a positive one. Two respondents (5.4%), observed cultural hindrances. According to them the community expressed concerns that women are not supposed to go out to work.

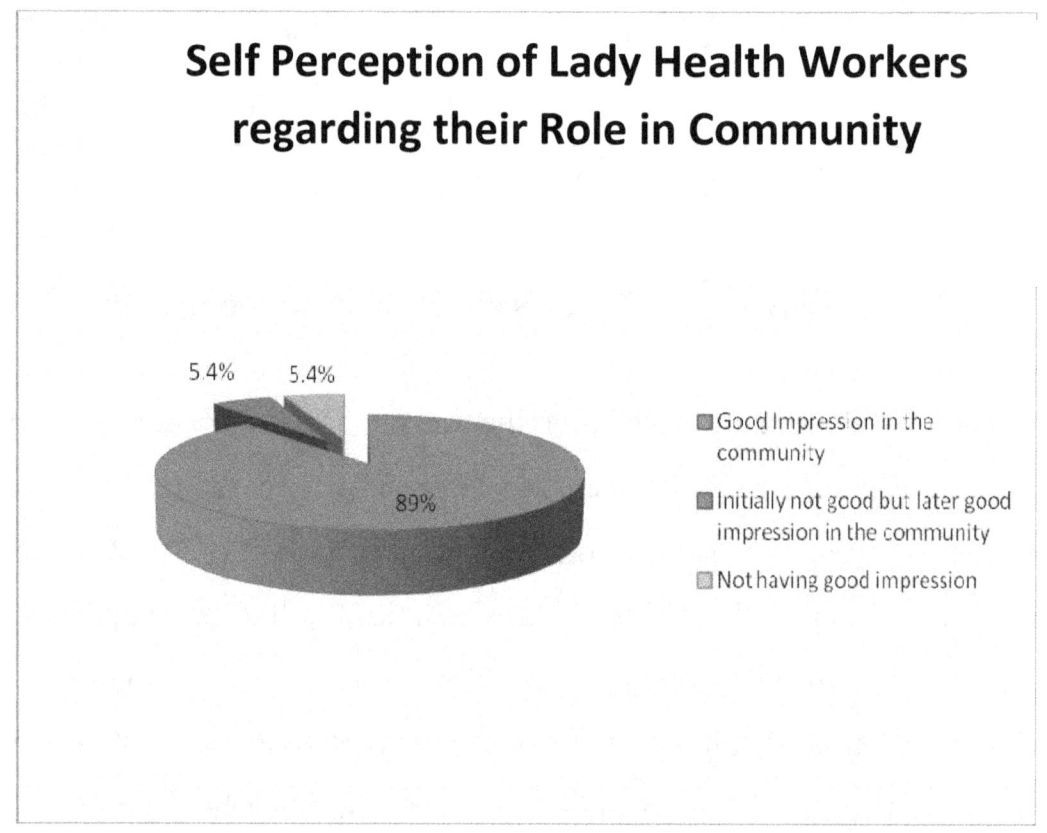

Figure: 1 Self perception of Lady Health workers regarding their role in community

S/No	Variable	Description
1	Polio campaign	60 %
2	Arranging meeting of Health Committee	40%
3	Attended trainings without logistic support	40%

Table: 5 Difficult Tasks during Service

Table: 5 demonstrate most common causes for LHWs getting overburdened and stressed out from their work, which ultimately forces them to resign. On the top were polio duties. About 60% considered polio duties as a difficult task and an additional burden, about 40 % considered arrangement of health committee meeting as a difficult task for them, especially with presence of male community members as interaction with males is considered to be a social & cultural stigma. It sometimes also causes hindrance in routine activities of the LHWs. Attending different trainings is also additional workload especially when appropriate logistic support is not provided and the incentives are very low.

In response to question related to the causes of LHWs resignation from the job. Table: 6 shows the actual causes of attrition of LHWs from the NP-FPPHC. Most of the LHWs gave multiple causes of their resignations, amongst which marriage and not giving time to children along with low salary compared to work load were the prime ones.

S.No	Variable	Frequency	Percentage
1	Domestic Reasons		70%
	a. Marriage	11	
	b. Not giving time to children/family	12	
	c. Transfer of husband	03	
2	Salary		60%
	a. Work is too much and salary is less	22	
	b. no logistic support		
3	Health Reasons		21%
	a. Cannot perform duty due to bad health	06	
	b. Old and cannot visit far flung areas any more.	02	
4	Problem with Supervisor		05%
	a. Not supportive	02	
	b. Always complaining against me		
5	Not having good impression in community		03%
	a. Objection by elders/ in laws	1	
	b. Culturally female job of roaming in community is not acceptable		

Table: 6 Causes of Attrition in Lady Health Workers of Pakistan

As aforementioned results from Table: 6 show that the attrition is mostly affected by the domestic reasons amongst which marriage and not giving time to children were on top. Inadequate pay was the second main cause of attrition. The results also demonstrate that LHWs complained of increased workload compared to which their salary was less, these LHWs also demanded logistic support for their duty. The salary of LHWs equals to 30 Euros/month, which is a very low amount to run the family. So mostly if they get better jobs they leave the LHWs job.

About 6 respondents cited medical reasons as a cause of resignation. Two Ex-LHWs were pregnant and advised rest during pregnancy. They complained that they were not allowed a long medical leave, so they had to resign. Old age associated with ill health was also reported as a cause of attrition. This reason is a matter of concern because these field workers are not given medical cover as a part of their job package, compared to rest of regular government employees. Only five percent of the LHWs had resigned due to problems with their supervisors. One LHW complained of lack of support by the supervisor, when the demands for her personal gains were not fulfilled, the LHS stressed her by change of behavior.

One LHW complained that her family members thought this job was not respectable, and will give a bad name to her and rest of the family, because of her community having a bad opinion about LHW's job, so she decided to resign.

In the questionnaire apart from the major questions related to attrition there were some other associating questions related to the programme which affect retention, attrition & job related satisfaction. These are tabulated in Table: 7.

S.No	Question	Y	N
1.	Did you felt empowered after becoming LHW?	24	13
2.	Did you face any problems during the initial three months training period?	8	29
3.	Were you satisfied by the introduction of initial three months training?	35	2
4.	Did you attend the monthly continued education sessions at the FLCF regularly?	34	3
5.	Was anybody from the FLCF staff present in those Sessions?	31	6
6.	Did LHS force you to do some personal favors?	1	36
7.	Has your health house ever been supervised/visited by any male member	17	20
8.	Did you ever had a negative experience with the male Supervisor?	10	27
9.	Were you comfortable with the community visits?	35	2
10.	Were you and your family comfortable with working in National Immunization Days?	16	22
11.	If you are given the opportunity to join again as LHW will you join again?	13	24

Table: 7 Job satisfaction of LHWs

While answering to the question related to empowerment the table above clearly explains that (64.86%) respondents felt empowered while doing their job as LHW. Similarly when asked about attending monthly continued education sessions 34 respondents said that they regularly attend the session while three were irregular in attendance. In response to the question of Lady Health Supervisor asking for some personal favors, only one respondent had problems with her LHS, rest of 36 had no problems with their LHS. When asked about visit of male supervisor to health house, 20 respondents were never visited by their male supervisors, but rest of the 17 had experienced visits by male supervisor, ten of which also complained of negative experiences with the male supervisors visit. Regarding the question of being comfortable with community visits, 35 responses were positive the rest two were not happy with the visits. When asked about the opportunity to join the programme again as LHW, 13 respondents gave a positive response while 24 were of the opinion that they will not join the programme again.

In order to find other minor causes some probing questions were asked to get a true insight of what LHW think about their job, some of these questions are described as follows;

In response to questioning about benefits of becoming LHWs, the responses received were multiple, Almost 27 joined for money along with some other reasons, which were as follows:

- We joined the job to earn Money — 27
- We were looking for professional respect — 13
- The reason for joining was to Increase our social — 09
- We wanted to benefit our community — 05
- We wanted to gain knowledge about health issues — 05

Training is one of the important aspects of the programme which has a great impact on LHWs performance, when asked about trainer's behavior and demonstration of professionalism responses obtained were

as follows;
- Excellent behavior 21
- Good 08
- No Comment 08

When the same question was probed by asking about any impolite attitude, or some form of harassment the replies were as follows:
- Acceptable behavior 29
- Impolite 05
- Reserved comments 03

(In privacy they mentioned that they had problems with their trainers for reasons they would not like to disclose).

On Inquiry why LHWs appreciated the monthly education sessions the response received were as follows;
- We receive training on some topic 33
- We can meet our supervisor 18
- Report correction is done 28

Similarly on inquiry about the part of their job they liked most Opinions of LHWs were as;
- We like field visits the most 25
- Training is the best part 15
- We can help the needy 20
- We can learn practical skills 04

To assess number of supervisor's visits and its contents the answers to queries were as follows;
- We get visited by LHS once a month 12
- We get visited by LHS twice per month 25
- During the visit the LHS checks all 17
- The LHS checks health house completely 11
- The LHS visits our community 10
- The LHS checks our practical skills 05

The questions related to the visit of male supervisors are already tabulated but further probing on what their family perceives about male visits, most respondents mentioned that these visits were not liked by their male relatives for the fear of being stigmatized.

As a part of the survey, interviews with the officers of the LHWs programme were also conducted. The idea was twofold, one was to countercheck some of the questions asked in the LHWs interview, and other was to know their opinion about the LHWs attrition and how it affects the programme. The following officers were interviewed:

- Provincial coordinator (PC),
- Assistant provincial coordinator (APC),
- District coordinators (DC) of the two study districts,
- Two Lady Health supervisors (LHSs), one from each district.

The preliminary questions in the questionnaire were related to the basic information of the concerned officer, the posts are already mentioned and the work experience of the officers on average was 4 years. The officers interviewed were working both on administrative posts, as well as involved in monitoring, supervision and field visits. The field activities for PC and APC were less frequent as compared to DC. The LHSs were mostly involved in field visits of LHWs.

In response to inquiries about the level of job satisfaction of LHWs, all officers interviewed were of the opinion that they are mostly satisfied, but some of them are definitely not. The reasons specified for low satisfaction levels were low salary, and cultural unacceptability.

Regarding the causes of LHWs resignation, all the interviewees were of the opinion that reasons like marriage, getting a better job, low salary and rarely illness are among the common causes, and in present scenario deteriorating security situation has become an additional cause.

All officers were concerned that even an attrition of less that 5% affects the programme quality; resulting in wastage of resources and trained workers. The area from where the LHW resigns is left deprived of any primary health care (PHC) services. This exerts a negative impact on programme itself, and creates problems for the officers in terms of negative political pressure, for purpose of inducting new LHWs. The commonest proposal as a remedial measure to reduce attrition was increase in monetary incentives and improving quality of training.

Chapter: 5

ARGUMENT

The Lady Health Worker programme is an increasingly important element in the plan of Government of Pakistan to improve the health status of its population, both in the rural areas and urban slums. The programme was started in 1994, it has expanded very rapidly and currently within a period of 18 years over eighty million people of the country are receiving services of the LHWs, exerting a positive impact on health status of the people (OPM 2009). There are a total of 89,000 LHWs in all four provinces of the country giving a net coverage of 63% to the whole population. In Khyber Pakhtunkhwa the number of LHWs is 12,500 covering 55% of the population in a total of 24 districts.

As compared to Pakistan where the number of LHWs is exceeding 89,000 workers studies done in other countries demonstrate they have lesser numbers of community workers. Most of these countries have volunteer based community programmes where instead of regular salaries, workers are getting nonmonetary incentives. Also, reports from different countries show a very high attrition rate (attrition rate of up to 60% has been reported), whereas in Pakistan the attrition rate was reported to be 5% (OPM 2003). Looking at the impact of work these workers have on healthcare delivery, countries including Pakistan are working hard to improve workers retention, as improving indices of worker's retention is necessary to maintain good access to health services.

The primary objective of this work was to get an insight into the reasons why the LHWs resign from the NPFPHC. The key findings which emerged can be summarized as follows:

- Domestic reasons like (marriage, less time for children, and transfer of husband) were the most common causes leading to attrition.
- Inadequate salary and too much work load (mostly due to National Immunization days).
- Ill health, bad behavior of the supervisor & perception of poor respect by the community were among other causes that lead to attrition.

These key findings are elaborated in detail with focus on how to control factors causing attrition as follows;

5.1 Importance of Knowledge and responsibilities

The survey findings of LHWs being aware of their job description was very encouraging, because before conducting this work, it was feared that LHWs have a very low education level, and may not be aware of their exact role & duties related to the field. The results showed more than 95% were fully aware about role and responsibilities related to their job. However due to low education level it is necessary to refresh the knowledge of LHWs regularly. For this reason the programme has a specially designed training system which is known as *"continued education session"*. The LHWs attend their respective health center one day each month to get refresher training on the topic identified. In addition to continuing education sessions, every year there is a 15 days refresher training in key areas. This survey also reflects feedback given by third external evaluation of the programme which has reported that "good quality training leads to improved knowledge which results in better performing LHWs and their supervisors" (OPM 2003).

5.2 Domestic reasons influencing attrition

Inferences show that marriage is a common cause for LHWs to resign. It was noticed that majority of LHWs joined the job at a young age (32 LHWs joined at the age below 30 years), similarly the ratio of unmarried LHWs at the time of joining was higher, but later they got married and resigned from job. Marriage as a common cause of attrition is consistent with the findings of studies from Solomon Island, where it is reported that the problem of attrition is common with young age. According to them, young workers quit the job when they get married and look for better jobs to support their families (Chevalier etal 1993). This statement about Solomon Island clearly defines that young age is not a good criteria for LHWs recruitment. Another study cited the reasons for attrition as movement to higher positions in health system, marriage or family matters, and finding better positions in other fields (Ofosu-Amaah 1983). Some other countries report the same problem as a common cause. These countries are trying different approaches for remedial of this cause of attrition, as in Bangladesh the BRAC programme which is a community based programme has implemented a policy to select only married and widow women as a community workers which can reduce attrition due to marriage (UNICEF 2004). Trying different ideas to improve retention of the workers are necessary to improve access & delivery of healthcare.

5.3 Attrition related to inadequate salary & lack of incentives

Our findings confirm low salary as another cause of attrition, this being consistent with findings of studies conducted in "Nigeria on Community Directed Distributors of Onchocerciasis control programme where the attrition due to lack of incentives was reported to be 65.9 %" (Annals Jan 2008). Studies show that low salary is a very important factor for worker's attrition, and in order to sustain the programme it is important to have well designed policies with introduction of some performance based incentives, in order to improve worker's motivation levels and to reduce dissatisfaction caused by low salary.

The problem of lack of incentives as a cause leading to attrition has been reported by most of countries having community based programme, like Bhutan, Nepal, Bangladesh, Nigeria etc .The reason for attrition, in these countries being a trend of volunteer based programmes with little incentives by the government during for training of community workers. The rest of income is generated through sales of medicines in the community.

Unlike above mentioned countries, Pakistan has a well established LHWs programme which is owned by the Ministry of Health Pakistan, and salaries are paid by the government. Inferences of this survey confirmed that inadequate salary is the second most common cause leading to attrition, 22 out of 37 respondents reported this problem as a reason for them to resign. However for another 15 LHWs it was not a major concern. When probing questions were asked; the responses were; that the workload of different duties allocated is much more as compared to salary given, the officers' interview also revealed the same response. They were of the opinion that the workload exerted by additional duties is much more compared to the salary. Although the salaries paid to the LHWs are low in amount, during the last five years the government has increased it from 1600 rupees/month to 3090 rupees/month. Increasing the salary on workers' demand is a difficult problem to address, as a very small increase generates a huge financial burden & the LHWs programme already consumes a large proportion of country health budget. However the fact being that when pay does not meet the employee need, they are compelled to find better paid jobs, resulting in high attrition of programme.

It was noticed that the attrition rate of LHWs programme Pakistan is less as compared to the other countries as evident from available literature. The reason for this could be that this is not a volunteer based programme as opposed to most of the countries and it has a regular salaries disbursement mechanism by the government itself. It is evident that CHWs who receive a government salary retain to their jobs more as compared to the ones who depend on community financing (UNICEF 2004). However monetary incentive can often create problems like, money may not be enough, may not be paid regularly or may stop suddenly as was seen in India and Nepal. Hence to improve the workers motivation level, it was suggested that instead of giving money, nonmonetary incentives (bicycle etc) can be tried to improve workers motivation. In addition giving role of curative services to these workers can also improve workers' retention (UNICEF 2004).

5.4 Health versus attrition

Health related problems had caused some LHWs to resign. A total of 6 LHWs reported bad health as a cause of their resignation. The LHWs programme is designed for difficult terrain, irrespective of the area where the LHW lives; she has to visit each and every house hold of her community. Her own ill health can definitely affect these field visits. The programme does not offer medical services as part of job package to these already low paid employees. They have to work even when they are ill, so they prefer to resign.

The same issue was identified in other countries as well, mentioning that; for better working conditions, factors affecting women workers should be critically analyzed, such as physical workloads, health and reconciling work & family (Gupta & et al 2006).

5.5 Behavioral factor

The fourth cause evident from Inferences was the problems with the Lady Health Supervisor's (LHS) behavior. 5% of LHWs reported unsupportive attitude of their supervisors. The problem of behavior affecting workers performance is a reported reason for attrition in various fields across the globe. Psychological violence or verbal abuse is very common in the workplace. Regular monitoring and supervision of environment that creates risks for workers is necessary. It also needs intervention for improvement (Jackson M & et al 2005). It is widely acknowledged and emphasized, that the success of Community health worker programmes depends on regular and reliable supervision (Ofosu-Amaah, 1983; Bhattacharya etal 2001).

5.6 Importance of community opinion

This issue was reported by only 3% of the participants of survey; who mentioned this job to be unrespectable culturally, moreover family members also had reservations related to involvement of fieldwork as part of job. Resistance from spouse has been identified as a key barrier to the participation of women in these programmes. (Brown et al 2006).

5.7 Self perception of LHWs

This work is the first time that an effort has been made to highlight importance of self perception of LHWs regarding their role in community. The fact that 89.1% of LHWs perceived themselves as being accepted by community is important for a successful programme. Evidence reflects that increased respect from the community and colleagues is a motivating factor. For retention of workers; in addition to incentives, community recognition and public appreciation is also necessary (Government of Nepal 2003). However in case Pakistan social unacceptability can be due to social setup of being a male dominated society where idea of females visiting other households is not welcomed by many.

5.8 Motivating versus discouraging factors

Main motivating factors for LHWs joining this job was to earn money, gain respect, increase social freedom, have knowledge about problems related to health and benefit their community from the same. However these reasons can later become a cause of their resignation as aforementioned.

5.9 Problems related to training

Training is one of the most important aspects of the LHWs programme with the problem of impolite attitude of trainers being a contributor to attrition. This issue of is consistent with what is reported in evidence worldwide; as community workers sometimes face problems at hands of male trainers. (NPFPPHC 2009), other reported problems of physical & psychological violence and verbal abuse are common in workplace. It is also reported that CHWs receive little support in terms of training management and supervision, resulting in dropout of volunteers due to lack of motivation (Hilary stand etal 2008). The document review during the study confirmed the presence of complaint letters against male trainers, which were further verified during the course of LHW's interviews.

5.10 Some aspects related to job satisfaction

Another finding of this survey revealed perception of LHWs about their jobs. Most of the LHWs were of the opinion that best and most liked part of their job were community visits. They mentioned that during community visits they get a chance to socialize along with opportunity to help their respective community for problems related to healthcare. This finding is consistent with evidence from elsewhere where authors reflect the role of health worker as an advocate for the community and an agent of social change, who works against inequities and is an advocate of community rights (UNICEF 2004). Training and a chance to learn practical skills is also appreciated by LHWs. Moreover it was revealed that most of the LHWs were comfortable with their jobs and were happy to perform field visits. The family problems related to National Immunization days were bothersome for some. Overall LHWs seemed quite satisfied with the nature of their job. To improve the LHWs effectiveness it is necessary to create a more conducive workplace environment for them by a strong health system support.

5.11 Problems related to male supervisory visits

Male supervisors visiting the health house is a common practice. LHWs mentioned couple of problems related to their visit; firstly it was reported that it creates a bad impression for them in the community; secondly it was also not liked by the male members of their family who feared stigmatization by community. This problem is of concern because the supervision is mostly done by the female LHSs, but sometimes to countercheck their visits it is necessary that the DC and other officers from the PPIU visit the LHW health house. If this practice is stopped completely, it will definitely affect the overall programme effectiveness.

5.12 Importance of attrition

The Lady Health workers programme plays an important role in empowering women by providing employment opportunities to them. This programme in remote areas, has also motivated communities to educate their girls, so that they can work as LHWs in those areas later on. After putting so many efforts, it is matter of dire importance to retain LHWs inducted in this programme. Attrition not only results in discontinuation of experience and further development of skills but also results in loss of trained personnel. In order to check attrition it is important to have more accurate data with magnitude of turnover to make necessary interventions. However before any interventions the following five points for workers retention need to be ensured;

i. Strong support by co-workers,
ii. Support from the community,
iii. Support by the family,
iv. Ensure regular basic training,
v. Availability of relevant staff.

While conducting this survey a discrepancy in records was noticed. The records were not well maintained in some of districts; also the flow of information from the district to PPIU was not proper, as was noticed during document review and reported by the PPIU staff. This results in misleading information and inappropriate reporting of actual figures giving misleading results.

Most of the reasons for attrition inferred from this survey cannot be directly controlled by the programme management like; domestic problems (marriage not giving time to children etc), ill health, not having good impression in the community. However some of them can be managed indirectly by formulation of proper policies with a transparent selection process and a thorough demographic investigation before induction of personnel. For improving ownership and acceptability by the community it is necessary to involve community opinion in selection process and also to ensure community involvement in various activities carried out by the LHWs. The model for this being collection of district level endowment funds to support the community workers. The funds are later transferred to the Village Development Committee (VDC) level endowment fund, which creates interest and encourages the VDCs to sustain these community workers" (UNICEF 2004).

CHAPTER: 6

Key Recommendations

Through this work it was possible to find out the causes of LHW's resignation in two selected districts of Khyber Pakhtunkhwa, Pakistan, contributing to overall attrition of the LHWs programme. The survey results revealed that most of the LHWs resign due to personal causes, secondly low salary with overburden of duties and the thirdly main findings were ill health, bad behavior of the supervisor and a poor respect by the community as main causes of attrition. The issue of attrition as mentioned above result in wastage of resources which are meant for other programme activities; leaving behind a lot of problems for the programme authorities. All the reasons related to attrition cannot be handled directly by the programme and its managers but some of them can be tackled by bringing minor changes in existing policies. Some key recommendations which can help in reducing attrition due to different causes are given below;

6.1 Policy formulation for special cases

For attrition due to marriage or transfer,

a) Preference should be given to induct already married females with a written consent from the family, so that this problem should not arise after few years of working.

b) To take some money as security at the time of appointment and signing a bond for 5 years of service especially for more educated ones.

c) Keeping strict liaison with the rest of health department, for adjustment of those LHWs who left the job due to marriage or shifting of home, these LHWs should be readjusted either by the programme itself if possible, or in other community based jobs of the health department (like Population Welfare Department or Maternal Neonatal Health programme etc).

6.2 Enhancement in salary

Either there should be a regular increase in salary or a reward based bonus system with the help of partner organizations. This is to help in compensating for low salary. Also, the extra work on National Immunization days can be additionally salaried or rewarded.

6.3 Provision of Medical facilities

With regards to LHW health services problem, medical services need to be provided to the LHWs as part of their job package just as the case is with rest of the staff of health department. Annual medical leaves and medical allowances which are provided to the rest of the health staff should also be provided to the LHWs. These measures should be made part of programme policy.

6.4 Formulation of incentives for additional duties

For the purpose of overburdening due to increased workload, additional duties assigned to LHWs like polio days etc should not be a compulsion for them, rather they could be taken as optional duty with additional pay along with incentives like certificates, monetary incentives and awards for the best performers.

6.5 Behavioral change by Supervisors

Strict monitoring and investigation system needs to be maintained from the office of District Coordinator and the PPIU managers, in order to check any disparity in attitude of the Lady Health Supervisors (LHS), In addition regular training on supportive supervision along with training on interpersonal communication should be delivered to the LHSs in order to make the working environment more conducive for the LHWs.

6.6 Improvement in record maintenance

It was observed that sometimes records were not being maintained properly. This can result in loss of important information. Therefore it is important to improve record keeping on LHW Characteristics and analyze it on regular basis to identify trends in attrition rates and demographics.

6.7 Creating impact through ownership

To deal with the issue of poor respect there should be ownership for the LHW programme by the health department.

a) The referred cases of the LHWs should be acknowledged at all levels of health care especially secondary and tertiary care hospitals, for this reason orientation meetings should be arranged with all relevant health staff.

b) Use of the ID cards should be made mandatory for LHWs, for proving their identity when needed.

c) Community opinion should be involved at the time of initial selection of LHWs.

6.8 Monitoring the trainings

LHWs face various problems during their regular trainings. This is a matter of concern and in order to solve this issue; programme authorities need to ensure regular monitoring of LHWs training from higher hierarchy involving the PC, DPC, APC, training coordinator and field programme officers. This is to ensure that LHWs feel supported, safe and secure during their training period.

The recommendations if followed will help programme authorities to design policies in a way which are helpful in retention of Lady Health workers of the programme. Thereby, preventing wastage of extra resources spent for new inductions. These resources can then be utilized for the betterment of the programme itself and result in improved functioning of the programme.

References

1. Annals of Tropical Medicine and Parasitology, Volume 102, Number 1, January 2008, pp. 45-51(7)

2. Berman PA 1984.village health workers in java, Indonesia: coverage and equity. Soc sci Med, 19 (4)411 – 422

3. BRAC, Health programmes http://www.brac.net/health.html.

4. Bhattacharyya K, Winch P, Le Ban K, Tien M. Community health workers incentives and disincentive: how they affect motivation, retention and sustainability. Published by the basic support for institutionalizing child survival project (basics 11) for the United States agency for International development 2001)

5. Bhattacharyya K, Winch P, Leban K, Tien M (2001). Community health workers incentives and disincentives: how they affect motivation, retention and sustainability. Arlington, Virginia, BASICS/ USAID

6. Brown A, Malca R, Zumaran A, Miranda JJ (2006). On the front line of primary health care: the profile of community health workers in rural Quenchua communities in Peru. Hum Resour Health 4(1):11.

7. Cameron, R. *Health Human Resources Trends in the Americas: Evidence for Action*. Pan American Health Organization, September 2006

8. Chatterjee, M. Health for too many: India experiments with truth. Reaching health for all (Edited by Jon Rohde, Meera Chatterjee and David Morley), pp 342-376. Oxford University Press, Delhi, 1993

9. Chevalier C, Lapo A, O'Brien J, Wierzba TF (1993). Why do village health

workers dropout? World Health forum, 14(3):258-261.

10. Columbia University site for Map
www.columbia.edu/.../00maplinks/.../pakphysical.html

11. Danida, Danish support to the health sector in Bhutan. Annex XIII: Information, education and communication for health and community participation, 2001. Available from http://www.un.dk/danida/evalueringsapporter/1999-10/bhutan/b13.asp

12. Dictionary US department of Defense dictionary of military and associated terms 2005, www.dtic.mil/doctrine/jel/new_pubs/jbl.02.pdf

13. Ebrahim G (1988).Learning from doing: progression to primary health care within a national programme. A case study from Tanzania Journal of Tropical Pediatrics, 34(1):4 – 11.

14. Farmer P, Leandre F, Mukherjee J, Gupta R, Tarter L, Kim JY(2001).Community-based treatment of advanced HIV disease: introducing DOT-HAART(directly observed therapy with highly active antiretroviral therapy).Bull world Health organ,79(12): 1145-1151.

15. Gilson L, Walt G, Heggenhougen K, Owuor-Omondi L, Perera M, Ross D et al, and National community health workers programs: How they can be strengthened? Journal of Public health policy, 1989:518-32

16. Gilroy KE, Winch (2006).Management of sick children by community health workers. Intervention models and programme examples. Geneva. WHO/ UNICEF

17. Gilson L, Walt G, Heggenhougen K, Owuor-Omondi L, Perera M, and Ross D, Salazar L (1989) .National community health workers program: How they can be strengthened? J public health policy, 10(4): 518 532.

18. Government of Nepal and Maternal and Neonatal Health, a study on concept of

volunteerism: focus on community-based health volunteers in selected areas of Nepal, 2003.

19. Harriet Burn July 2008, an insight into the reasons why Lady Health Workers resign from Pakistan's National programme for family Planning and Primary health care.

20. Hossain MM. BRAC's Shastho Shebika: A case study of community health volunteers approach in Bangladesh, 1999.

21. Information obtained by researcher through NP-FPPHC Administration office, keeping records of HR reports Peshawar main office

22. Interim Poverty reduction strategy paper Pakistan January 31st 2002
www.imf.org/external/NP/prsp/prsp

23. International database demographics of Pakistan world population data sheet population reference Bureau 2009
en.wikipedia.org/wiki/Demographics_of_Pakistan

24. Jackson M, Ashley D. Physical and psychological violence in Jamaica's health sector. Rev Panam Salud Publica. 2005; 18 (2):114-121

25. Khan SH, Chaudhury AM, Karim F, Barua MK (1998). Training and retaining Shasthyo Shebika: reasons for turnover of community health workers in Bangladesh Health care Supervisor, 17(1):37-47.

26. Kaufmann KS, Myers DH (1997). The changing role of village health volunteers in northeast Thailand: an ethnographic field study. Int J Nurs Stud: 34(4):249- 255.

27. Lancet 2004: As part of the Joint Learning Initiative in 2004, a 117 cross-country econometric study was conduct that concluded that density of health

human resources is significant in accounting for maternal mortality, infant mortality and under-five mortality rates, in addition to and independent of policies that bring about income growth, poverty reduction and increases in female education. See: Anand S. Barnighausen T. Human resource and health outcomes: cross-country econometric study. The Lancet 2004; Vol 364: pp. 1603-09.lume

28. Lewin SA, Dick J, Pond P, Zwarenstein M, Aja G, van Wyk B, Bosch-Capblanch X, Patrick M (2005). Lay health workers in primary and community healthcare. Cochrane Database System Rev (1):CD004015.

29. Lady health Workers Management Information System NP-FPPHC 2009, information obtained from MIS section Peshawar main office.

30. Maru, RM. The community health workers scheme in India: an evaluation Social Science and Medicine. 17(19),(1477-1483), 1983.

31. Ministry of health government of Pakistan: Internal assessments of lady health workers programme 2007, National Programme for Family Planning and Primary Healthcare Islamabad 2008.

32. Mullan F: The metrics of the physician brain drain. N Engl, J Med. 2005 353:1810–8. Doi: 10.1056/NEJMsa050004. [Pub Med]

33. National Programme for Family Planning and Primary Health Care Management Information System 2009

34. Narasimhan.V V, Brown H, Pablo's-Mendez A, Adams O, Dussault G, Elzinga G, et al. Responding to the global human resources crisis. Lancet.2004; 363:1469–72. Doi: 10.1016/S0140-6736(04)16108-4.

35. National Programme for Family Planning and Primary Healthcare PC-1 (2003 – 2008). Ministry of health government of Pakistan

36. National programme for FP & PHC PC-1 2003-2008

37. Neeru Gupta and Mario R Dal Poz, Human Resources *for Health* 2009, http://www.human-resources health.com.

38. NP-FPPHC 2008 reviewed data at EDOH, complaints against LHSs & FPOs

39. NPFPPHC: The definitions mentioned are obtained by researcher through NP-FPPHC Administration office, keeping records of HR reports Peshawar main office

40. Oxford Policy Management Lady Health Workers Programme External Evaluation of the National programme for Family planning and primary health care final report. Oxford 2003

41. Ofosu-Amaah V (1983). National experience in the use of community health workers a review of current issues and problems WHO offset Publ, 71:1 - 49

42. Oxford Policy Management, NP-FPPHC external evaluation, quantitative survey report 2002, & Primary Health Care wing Ministry of Health Pakistan 2006.

43. Oxford Policy Management, NP-FPPHC external evaluation (UK and funded by the Canadian International Development Agency through a World Bank Trust fund) 2009

44. PHC wing Ministry of Health. Annual report: National program for family planning and Primary health care August 2005-September 2006.2006, Government of Pakistan

45. Parlato M, Favin M (1982).Primary health care. Progress and problems: an analysis of 52 A.I.D. assisted projects Washington DC, American Public Health association

46. Pan American health organization, World health organization 140th session of Executive committee Washington D.C., USA, 25 – 29 June 2007

47. Poverty reduction strategy paper Government of Pakistan December 2003

48. Rubin PN. Civil rights and criminal justice: primer on sexual harassment. National institute of justice, October 1995

49. SOCHARA (2005) an external evaluative study of the state health resource centre (SHRC) and the Mitanin Programme final report Bangalore, Society for community health awareness. Research and Action (SOCHARA)

50. Sanders,1992; Sauerborn, Nougtara & Diesfeld, 1989; Twumasi & Freund, 1985; Streefland, 1990; Ebrahim, 1988.

51. Streefland PJC (1990).Implementing primary health care Amsterdam, royal tropical institute

52. Sringernyuang L, Hongvivatana T, Pradabmuk P (1995).Implications of community health workers distributing drugs a case study of Thailand. Geneva, World Health Organization

53. Sanders D (1985), the struggle for health Medicine and the politics of underdevelopment London, Macmillan

54. Sanders D (1992) the state of democratization in primary health care: community participation and the village health workers programme Sauerborn R, Nougtara A, & Diesfeld HJ (1989) Low utilization of community health workers: results from a household interview survey in Burkina Faso. Soc Sci Med. 29(10):1163-1174.

55. Sharma, H., UNICEF Bhutan. Personal communication, 6 August, 2003

56. UNICEF, what works for Children in South Asia report: Community Health Workers? UNICEF South Asia, 2004

57. UNICEF 1993/94 to promote and coordinate recruitment, selection, training and administrative staff: evaluation of UN volunteer programmes at www.unjiu.org/data/reports/2003-07.pdf

58. USAID report copy right 2008 John Snow, Inc.

59. United Nations Millennium Development Goals 2009[Home page]. Available from: http://www.un.org/millenniumgoals/

60. WHO (1989).Strengthening the performance of community health workers in primary health care Report of a WHO study group Geneva, World Health organization (WHO Technical report series, No.780)

61. WHO (1990).Strengthening the performance of community health workers Geneva World Health Organization

62. Walt G (1992) Community health workers in National programmes Just another pair of hands? Milton Keynes Open University Press

63. Walt, G et al Are large-scale volunteer community health worker programmes feasible? The case study of Sri Lanka Social science & Medicine 29(5): 599-608, 1989

64. Waterston T, Sanders D (1987). Teaching primary health care some lessons from Zimbabwe Medical education, 21:4-9

65. Werner D (1981) the village health worker – lackey or liberator World health forum, 2:46-54

66. WHO report 2005 WHO evidence and information for policy, Department of human resources for health Geneva, January 200712.

67. WHO evidence and information for policy, Department of human resources for health Geneva, January 2007

68. World Health report 2006: working together for health Geneva: World Health Organization; 2006

69. Yahoo world news: http://www.dnaindia.com/world/report_unhcr-official-shot-dead-in-pakistan_1274498, Thursday, July 16, 2009 12:45.

70. Zurn P, Dolea C, Stilwell B. Nurse retention and recruitment: developing a motivated workforce [Issue paper 4]. Geneva: International Council of Nurses; 2005. Available from: http://www.icn.ch/global/Issue4Retention.pdf

ACKNOWLEDGEMENTS

This work is by no means a result of an individual effort of authors. A number of people contributed to this work directly or indirectly .We would like to thank Mr. Rainer Kuelker for his constructive guidance and supportive advice, The entire administration and staff of the Department of Tropical Hygiene and Public Health of the University of Heidelberg for their support. Dr. Haroon Rasheed Afridi for his technical guidance during the survey.

We would also like to especially thank the entire team of National programme for FP and PHC, in particular Dr. Ihsan Turabi, Dr. Muhammad Atif, Mr. Kaleem- ur-Rehman, Mr. Qazi Munir, Mr. Arshad Kaleem, Mr. Muhammad Imran, GTZ Peshawar Pakistan particularly Mr. Paul for their generous support and technical guidance.

We are especially thankful to District coordinators & Lady Health Supervisors of Districts Mardan and Nowshera, for their efforts in taking this work to field despite of ongoing security problems.

Finally we are extremely grateful to our family members for their support that enabled us to fulfill our pursuit.

Romana Ayub & Saad Siddiqui

ANNEXES

Annex: 1 Map of Pakistan Showing Khyber Pakhtunkhwa (Former NWFP)

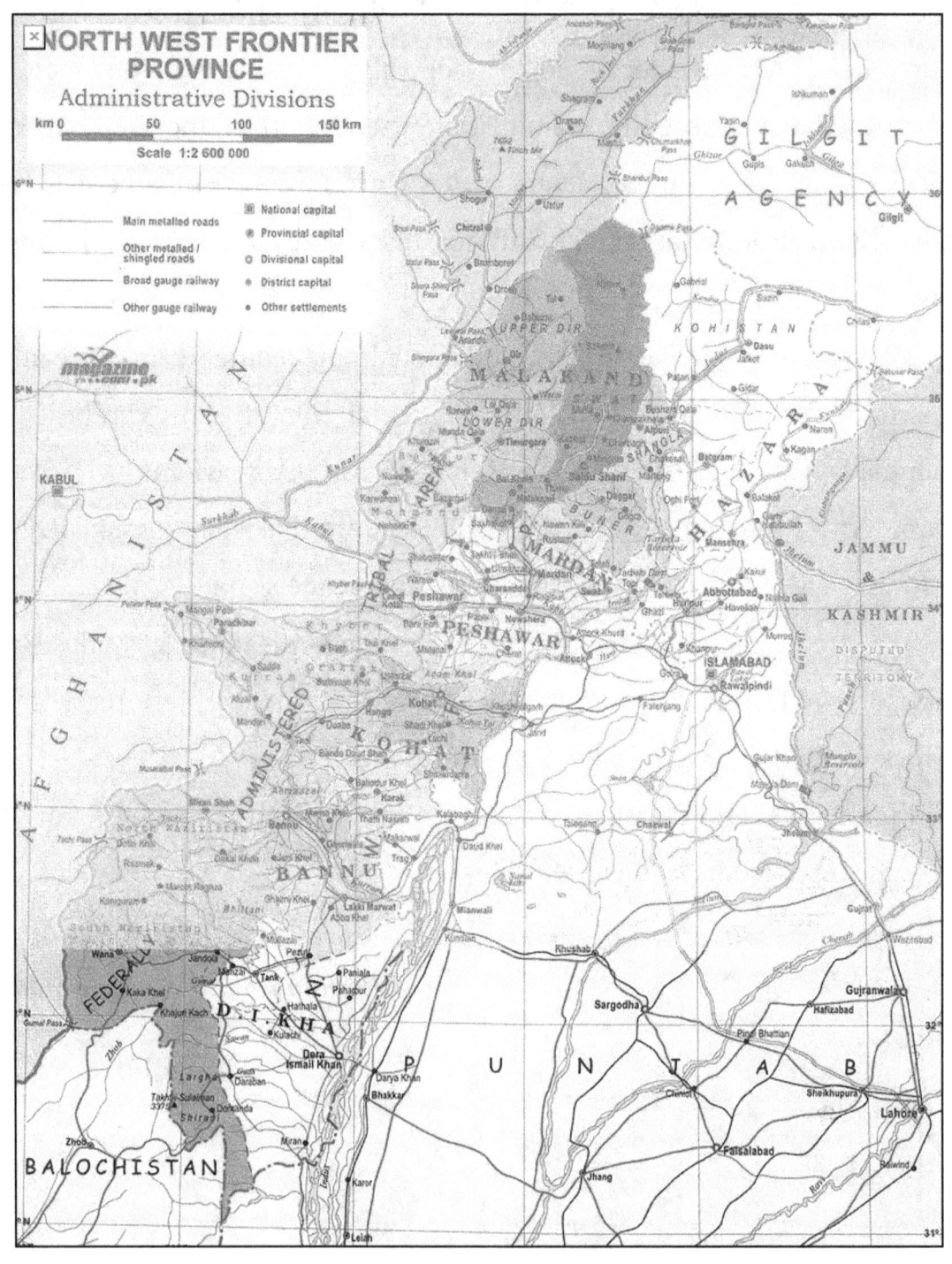

"Columbia University 2007"

Annex: 2 Study Districts Profile

Nowshera

Starting Year of Program	1995
Area of the district	1748.Sq kms
Total Population	1.05 M
Rural	1036170
Urban	18455
Covered by LHWs	735116
Rural	723116
Urban	12000
Total No. Health Facilities	57
Involved in National program	37
Working LHWs	775
Working LHSs	30

Mardan

Starting Year of Program	1995
Area of the district	1632 Sq; kms
Total Population	1969011
Rural	1378308
Urban	590703
Covered by LHWs	1176131
Rural	1034810
Urban	141321
Total No. Health Facilities	77
Involved in National program	60
Working LHWs	1225
Working LHSs	46

"National Programme records 2008"

Annex: 3 Document review tool

S.No	Reason of review	Document reviewed
1.	To count total attrition of LHWs, then filter the resigned ones from it	PPIU and DPIU records
2.	To confirm deletion of their names from the necessary records	Updated status of district monthly reports, payroll and the main register
3.	To check the details LHWs in district office like explanation calls, warning letters etc	Personal file
4.	To see weaknesses in reporting of LHWs in if any explanation was called for that	Monthly reports for weaknesses reported
5.	To check deletion of names from the list of working LHWs	District Monthly reports, annual reports, and the payroll

Annex: 4 Ethical note

On one hand resignation of trained Lady Health workers is adversely affecting the performance of the program and also leads to the loss of resources of this poor country and on the other hand it deprives LHW's family of a regular earning. After exploring these reasons on scientific methods I will forward evidence based recommendations to Ministry of Health for remedial measures. This can be made possible only, when you answer all the questions to the best of your understanding, honestly and without any fear. This interview has nothing to do with any official business and extreme confidentiality will be ensured regarding the information given. The interview will take about 60 minutes.